...tions patterns between
...distance telephone calls
...inking the areas of the
...number of calls.
...outgoing international-
...reveal the strength of

... and from New York
... hanges.
... ows for the top fifty
cit... York City on any given day.

Taken together, these visualizations explore the global reach of New York City by revealing the richness encoded in AT&T's telecommunications data.

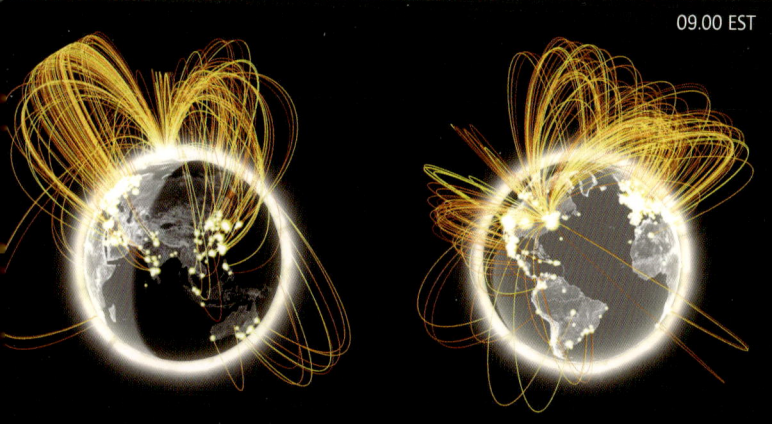

09.00 EST

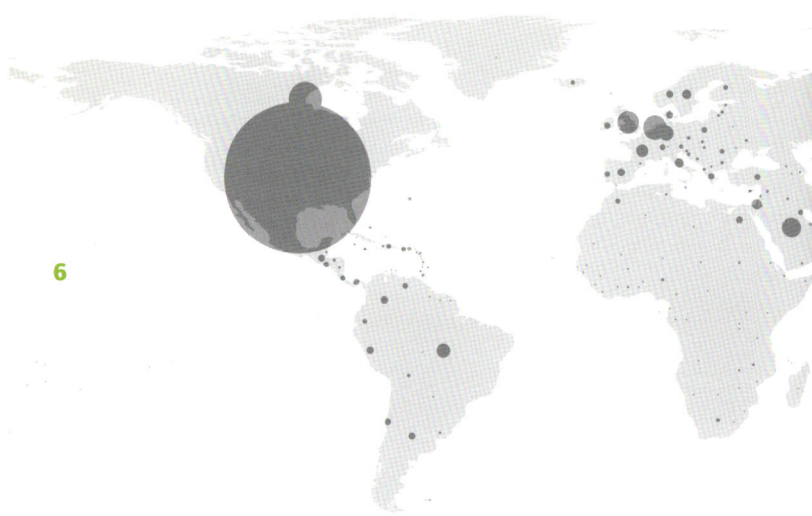

6

◀▲ 5-6
SKETCHES We used sketches such as these to explore the possibilities and range offered by the AT&T data. The sketches represent sample data for the total IP traffic flowing out of New York City to countries throughout the world.
Mercator map projection

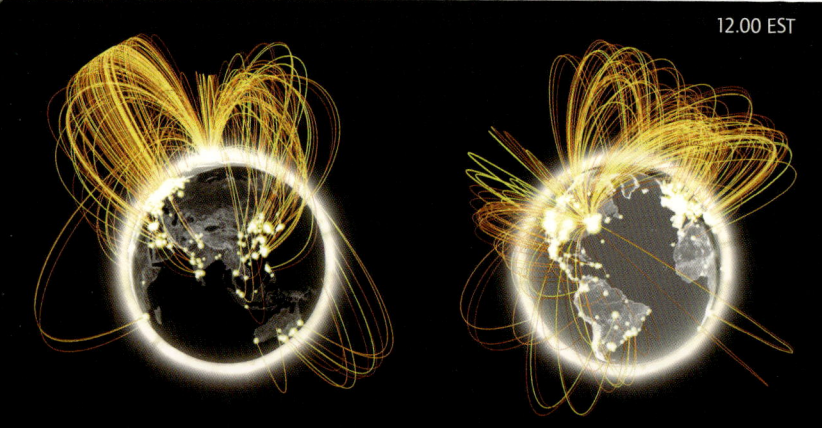

12.00 EST

▶ 8

CITY MAGNET This visualization compares all IP data flows with all international telephone calls from the city of New York. Connecting world cities are arranged sequentially around the circle according to their time zone, with Eastern Standard Time placed at the top of the circle. The red line represents voice calls while the yellow line corresponds to IP data flows. The length of each line is determined according to that city's percentage share of data or voice call exchange. Business cities such as Beijing and Toronto show greater flows of IP data than voice calls, whereas transnational immigrant cities such as Santo Domingo and Kingston have much stronger connections with New York through phone calls.

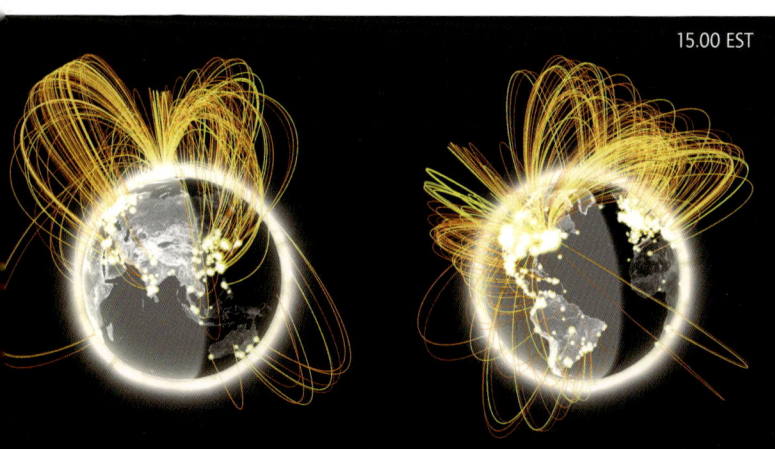

15.00 EST

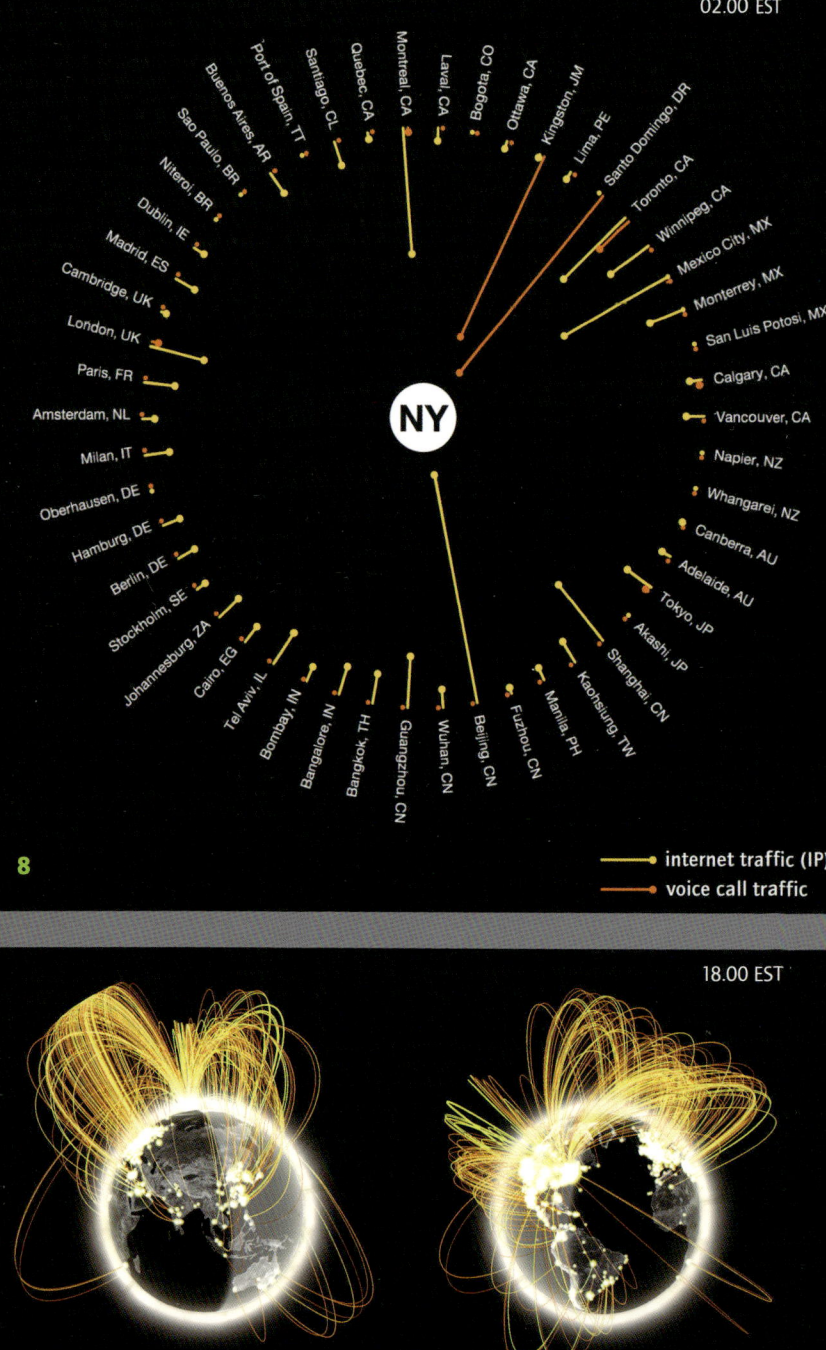

▼ 7
GLOBE ENCOUNTERS TIME SERIES

The globes in this time series are snapshots taken at 3-hour intervals on January 11, 2008. They represent the combined flows of IP data – such as email, peer-to-peer and web browsing – to and from New York City and other cities around the world. The size of the glow on a particular city location is proportional to the amount of IP traffic flowing between that place and New York City. Unlike the visualizations of international long-distance calls, however, Globe Encounters includes American cities in the visualization. Since nearly 80% of all the IP traffic on the AT&T network originates in the United States, Globe Encounters uses a higher threshold to produce the glow on American cities than on international cities.

As a result, the glow seen on west coast cities of the United States represents a larger amount of IP data than the similar glow on European cities.

▶ 9
NEW YORK P2P CITY RANKING
This table lists the top twenty-five foreign cities exchanging data with New York City through peer-to-peer networks. The cities are arranged according to their percentage share of the total peer-to-peer data flows observed during the week of January 8, 2008.

▶▶ 10
GLOBE ENCOUNTERS
This image of the Globe Encounters visualization represents the peer-to-peer exchange of IP data over AT&T's New York network on January 11, 2008. *Peer-to-peer* refers to data transfers between two computers where, in this case, one computer is in the New York metropolitan area (Connecticut, New York and New Jersey) and the other computer is somewhere else. An outgoing flow is a direct request for media content by a computer in New York to a computer in another city. An incoming flow refers to the delivery of media content from another city to the requesting computer in the New York network. The most well known type of peer-to-peer data transfer involves sharing of digital media such as music and movies.

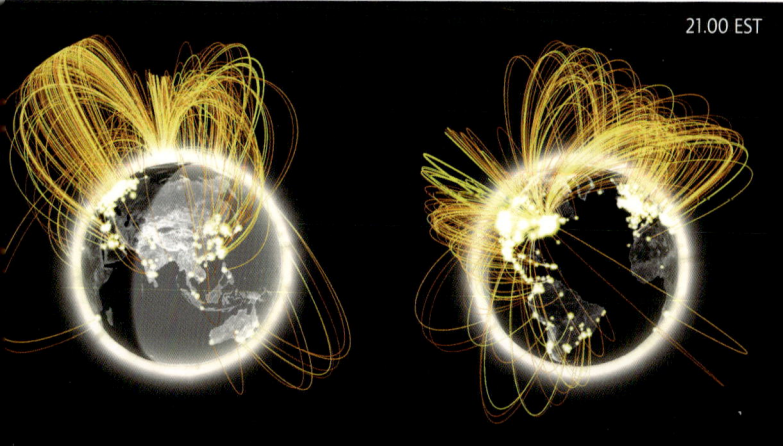

21.00 EST

City	Score
Seoul, South Korea	0.86
Beijing, China	0.85
Taipei, Taiwan	0.83
Riyadh, Saudi Arabia	0.68
Mexico City, Mexico	0.64
Bogotá, Colombia	0.47
Winnipeg, Canada	0.42
Rome, Italy	0.38
Shanghai, China	0.30
Athens, Greece	0.29
Buenos Aires, Argentina	0.26
Montreal, Canada	0.24
Toronto, Canada	0.23
London, United Kingdom	0.23
Taichung, Taiwan	0.22
Milan, Italy	0.22
Tel Aviv, Israel	0.19
Ramat Gan, Israel	0.18
Berlin, Germany	0.18
Moscow, Russian Fed.	0.17
Hamburg, Germany	0.17
Santiago, Chile	0.17
Sydney, Australia	0.17
Jeddah, Saudi Arabia	0.16
Monterrey, Mexico	0.16

NEW YORK P2P CITY RANKING

NYTE

visualizations

sense*able* city lab
Carlo Ratti
Kristian Kloeckl
Assaf Biderman
Francesco Calabrese
Margaret Ellen Haller
Aaron Koblin
Francisca M. Rojas
Andrea Vaccari

NYTE

essays

Carlo Ratti
Saskia Sassen
William J. Mitchell
Anthony M. Townsend
Mitchell L. Moss
AT&T Labs Research
Francesco Calabrese

MIT | sense*able* city lab:.::

Copyright
© 2008 by sense*able* city lab
All rights reserved.
Cambridge, Massachusetts

Published on the occasion
of the exhibition *Design
and the Elastic Mind*,
which ran from February 24
to May 12, 2008 at MoMA
The Museum of Modern Art
in New York City.
The exhibit was organized
by Paola Antonelli,
Curator and Patricia Juncosa
Vecchierini, Curatorial
Assistant, of MoMA's
Department of Architecture
and Design.

Project sponsor
AT&T

Technical partners
Yahoo! Design
Innovation Team
British Telecom

Under the patronage of
The Italian Cultural
Institute of New York

Produced by
The Massachusetts Institute
of Technology

Edited by
Francisca M. Rojas
Clelia Caldesi Valeri
Kristian Kloeckl
Carlo Ratti

Book concept and design by
Clelia Caldesi Valeri

Published by
SA+P Press

ISBN 978 0 9794774 1 6

Printed by
Aktiva in Turin, Italy

In reproducing the images
contained in this publication,
the sense*able* city lab
obtained the permission
of the rights holders whenever
possible. In those instances where
the sense*able* city lab could
not locate the rights holders
notwithstanding good-faith
efforts, it requests that any contact
information concerning such rights
holders be forwarded
so that they may be contacted
for future editions.

CONTENTS

7   New York Talk Exchange
by Carlo Ratti

10   New York City's Two Global Geographies of Talk
by Saskia Sassen

17   Mapping Urban Flux
by William J. Mitchell

25   New York: The City of the Telephone
by Anthony M. Townsend and Mitchell L. Moss

31   Representing AT&T's Network Data in Real Time
by Alexandre Gerber, Michael Merritt, Chris Rath and Jim Rowland of AT&T Labs Research

33   Data and System Architecture
by Francesco Calabrese

35   *Project Credits*

36   *Acknowledgments*

CARLO RATTI

NEW YORK TALK EXCHANGE

"Un bon croquis vaut mieux qu'un long discours."[1] (Napoleon).

Ironically, one of the greatest *croquis* ever drawn depicted one of Napoleon's greatest defeats. Charles Joseph Minard's classic *Carte of the Russian Campaign* of 1812-1815 condenses the story of that disastrous military adventure to just a few lines.[2] The passing of the seasons, the harshness of the climate, and more poignantly, the rising death tolls that decimated his troops are compellingly represented in one concise graph.

To a certain extent, Minard's map has been the inspiration for the graphic design of the New York Talk Exchange project on view at The Museum of Modern Art. As in Minard's case, we had to succinctly combine many types of information: New York's connections across the planet; the linkages between the city's different neighborhoods and other cities, countries and continents; and the dynamic changes that happen in the course of a day as time zones sweep across the globe. How to do it?

The source data provided to the sense*able* city lab by AT&T is based on countless entries in the following format:

Time of Day | Switch of Origin | Destination City | Traffic (megabyte data)

It should be noted that this telecommunications traffic is measured at a high level in the global network of flows. No information about

---

[1] "A good picture is worth more than a lengthy discourse."
[2] Minard's map was described by Edward Tufte as "the best statistical graph ever drawn."

individual calls or data transfers is collected at any point in the process, thus the end user's privacy is assured (for additional information on data collection, please see AT&T's data statement on p. 31). Considering the flow of bits at a planetary level is like looking at a river from a distance; it is not about tracing individual particles but about surveying the entire stream.

The seeming simplicity of AT&T's data format conceals some very complex dynamics that exist between New York and cities around the world. To reveal these complexities, we decided to create two different types of visualizations. The first one aims to show New York's global connections to the international network of cities – a kind of 'globalization in real time.' The second type of visualization zooms inside New York City's five boroughs and explores how global connections vary in different neighborhoods – a kind of 'globalization from the bottom.' As Saskia Sassen puts it (see page 10):

> Global talk happens largely among those at the top of the economy and at its lower end. This point is one of the striking pieces of evidence coming out of the data analyzed here. The vast middle layers of our society are far less global; the middle talks mostly nationally and locally…

The MoMA exhibition, and the illuminating essays contained in this book, mark the beginning of our work. Our plan in the coming months is to analyze the data in more depth. We hope to address some important research questions that loom behind the MoMA visualizations: how is the structure of global cities evolving? How could telecommunications data allow us to gain new insights into the dynamics of globalization? How do byte transfers across the globe affect the need for travel and physical displacement? The list goes on.

Other Talk Exchange projects might follow. The exciting aspect of NYTE is that it offers the potential to build a comparative database of global cities. British Telecom, for example, has shared data with us that show the top 100 locations connecting to London, arranged by IP traffic volumes for a day in December 2007. While the statistical significance of this data will have to be explored further, an initial comparison with equivalent IP data from the AT&T network reveals an intriguing dynamic. There has been much discourse in the popular media about the rivalry between New York and London for world city preeminence, with the majority concluding that London is the more cosmopolitan city.[3] However, the telecom-provider data reveals something different. Of the top 100 cities connecting with London – as determined by a comparison of IP traffic volume – all are located either in Europe or North America. In contrast, the top 100 cities connecting with New York include Asian and South American hubs such as Beijing, Bogotá and Riyadh. By examining the two carriers' data, we can surmise that London has a more limited 'cyber-hinterland' than New York. London seems to look to more established economies in the north, while New York opens itself up to emerging markets in the Middle and Far East and the global south. The AT&T and BT data comparison hints at a striking parallel: in an age of globalization, perhaps London's relationship to Europe is analogous to what is conventionally believed to be New York's relationship to the whole of the United States. The 'continent' may be closer to London than the British believe.

This cursory analysis illustrates how telecom data can help us to expand our conception of global cities and their role in the process of globalization. In the end, the NYTE project reveals as much about the city of New York as it does about its worldwide counterparts in immigration, business and culture. In other words, our visualizations demonstrate that in the information age, urban life is as global as it is local.

Carlo Ratti is Associate Professor of the Practice of Urban Technologies at the Massachusetts Institute of Technology. He directs the sense*able* city lab, a new research initiative that explores how technology is transforming urban design and living. He is also a founding member of the design office carlorattiassociati - walter nicolino & carlo ratti.

---

[3] "New York defines the metropolitan, London the cosmopolitan." Quoted from Harding J. 2007. "London Calling". *Times Online*, March 13, 2007.

SASKIA SASSEN

NEW YORK CITY'S TWO GLOBAL GEOGRAPHIES OF TALK

Two 24-hour geographies. Both are actually rolling, but one is the same actors as they move across the globe, the other is a geography of countries of origin, a roving talk machine that moves across the globe.

They capture globalization in action – talking.

Global talk happens largely among those at the top of the economy and at its lower end. This point is one of the striking pieces of evidence coming out of the data analyzed here. The vast middle layers of our society are far less global; the middle talks mostly nationally and locally, albeit in highly variable geographies.

Occuring at the top is increasingly, though not fully, a permanent twenty-four hours of talking, with rapidly shrinking 'nights.' This is the network of the forty or so global cities around the world where financial instruments are traded, new accounting models devised, mergers and acquisitions executed, and new ways of extracting profit invented. Traders today start at 04.00 or go on until midnight in some parts of the world so as to catch the end or the beginning of the day on the other side of the globe. The idea of the 24-hour financial center, awake and ready to trade with the whole world, took much longer to take shape than forecasters expected. In fact, it is still only a partial reality. But night-time as downtime is definitely a much shorter part of the 24-hour cycle than it used to be. And daytime as the time when all systems are going is definitely a brutally extended part of the cycle.

At the lower end, the 24-hour geography of global talk emerges out of the fact that the countries sending immigrants to New York circle the globe. As the Dominican Republic goes to sleep, Italy is about to wake up and so on across India and then the Philippines and China. If you call Manila from New York at midnight on Monday you will find them having lunch on Tuesday.

There are clearly also specific geographies that fit into neither one of the two major ones focused on here. For instance, calls between Jerusalem and Brooklyn and Queens are part of a more classic diasporic geography of communication. Secondly, calls between New York City and Geneva are part of the supranational system, with the United Nations headquartered in New York and the largest single concentration of U.N. agencies in Geneva. As with the major geographies of global talk focused on here, the AT&T data is a partial representation of all communication, given the proliferation of carriers in both originating and destination countries.

Even though global talk going out of and into New York City connects the city to multiple places worldwide, there is clear dominance of a few places. The AT&T data, only one of several carriers handling NYC's telephony, shows that London, Santo Domingo, Toronto and Kingston (Jamaica) are the main destinations for calls out of Manhattan. And the first three are also the largest originators of calls into Manhattan. Interestingly these top four already contain both global geographies of talk – one the world of transnational professionals (London and Toronto) and the other largely the world of immigrants.

Calls between Manhattan and London, Toronto, Tokyo, Hong Kong, Luxemburg, Singapore, Paris, Frankfurt, Zurich, Amsterdam, Shanghai, Madrid and Bangalore all constitute mostly the transnational professional global geography of talk, consisting of both foreign-born and native-born. There is a set of cities, notably São Paulo, Mexico City, Dublin and Mumbai which are likely to contain both this world and that of immigrants.

New York City's total foreign-born population stood at almost 2.9 million according to the 2000 U.S. Census. The largest single groups were 370,000 Dominicans, 262,000 Chinese and 179,000 Jamaicans. While their numbers may be small, they do a lot of the

NEW YORK CITY'S TWO GLOBAL GEOGRAPHIES OF TALK

The Paraculture, 2003
credit Hilary Koob-Sassen. T+2 Gallery, London

global talking out of and into New York City: they are the new high-level transnational professional class. Immigrants and transnational professionals, both foreign-born and native-born, are the two main global talking groups in NYC.

Each of these aggregated geographies of talk includes multiple differences and particularities. Returning to the case of cities that mix both our geographies of talk, it might be worthwhile to focus on Mumbai. The Indian population grew by thirty-three percent from 2000 to 2005 in the New York region (which includes New Jersey and Connecticut). This is one of the fastest growth rates; it now stands at well over 300,000. About a third of these reside in New York City proper. It is a very diverse population: there is an older professional class that includes university professors, a new professional class linked to global finance, a high-tech workforce and a very large group of small shopkeepers. Some of this can be caught from the talk data, but much cannot. The largest share of AT&T calls to India in NYC are between Mumbai and Manhattan; each of the major segments of the Indian population is probably included in these data. Yet Mumbai ranks 24th as a destination and 11th as an originator of calls into Manhattan; Mumbai is 32nd in calls into Brooklyn and 11th in calls into Queens. The exception to Mumbai's dominance among Indian cities is in Staten Island, where most calls to India are to Hyderabad. But let me also note that the top fifty cities worldwide calling into Staten Island do not include any from India. Given the size of the Indian presence in New York City, including many commuters from the suburbs outside NYC, this would seem to indicate that a good part of the calling is happening through other carriers. India has rapidly growing capabilities and a competitive market position to provide these global services.

Some of the AT&T data capture with astounding clarity particular geographies of talk. Thus on November 1, 2007, Kingston (Jamaica) accounted for about ten percent of all calls out of Brooklyn. Together, Kingston, Santo Domingo and Haiti (no city specified) account for seventeen percent of all calls out of Brooklyn. In the Bronx there is a symmetry between incoming and outgoing calls. Kingston and Santo Domingo accounted for thirty percent of all calls out of the Bronx. Santo Domingo and Santiago (the second major city in the Dominican Republic) accounted for almost twenty percent of the calls going into the Bronx. But there are also notable asymmetries: Toronto accounted for almost five percent of AT&T calls coming into the Bronx but only one percent of calls from the Bronx. Part of this asymmetry may have to do with different carriers. Both of these global geographies of talk never stop – it is 24-hour talking. But how that round-the-clock talk is constituted varies sharply. In one case it is a single articulated space, embedded largely in the new global economy. In the other it is a rolling wave that moves from one country to the other as one wakes and the other goes to sleep.

Saskia Sassen is the Lynd Professor of Sociology and Member of The Committee on Global Thought at Columbia University. Her recent books are *Territory, Authority, Rights: From Medieval to Global Assemblages* (Princeton University Press 2006) and *A Sociology of Globalization* (Norton 2007).

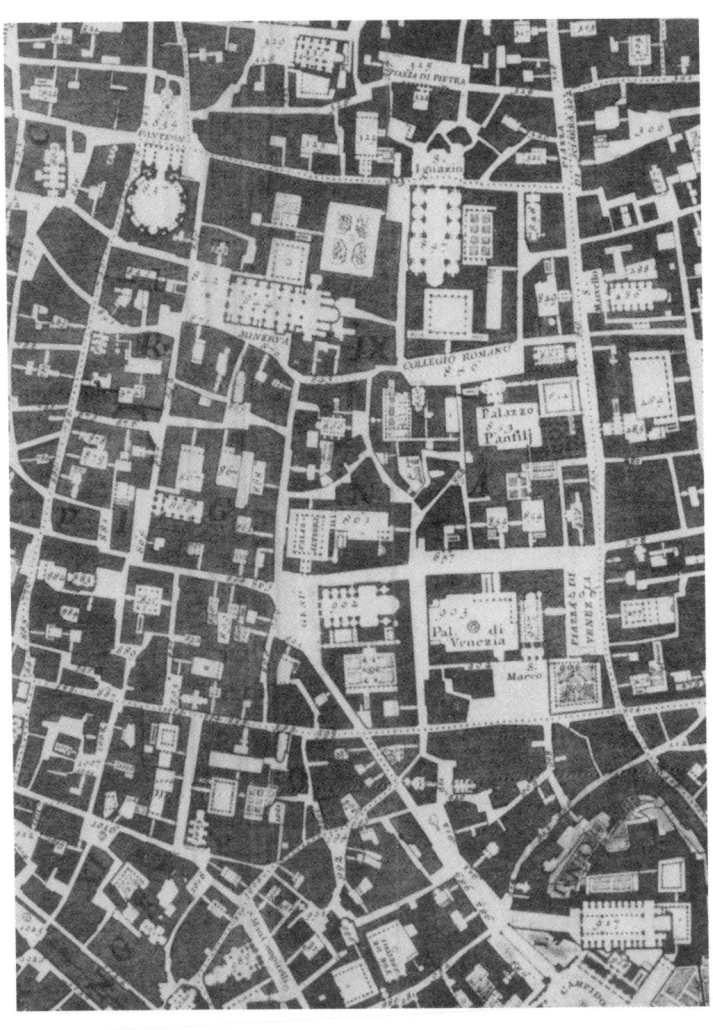

Nolli Map of Rome, 1748
credit Earth Sciences & Map Library at the University of California, Berkeley

WILLIAM J. MITCHELL

MAPPING URBAN FLUX

The city shown in Giambattista Nolli's great eighteenth-century map of Rome looks as if it has been hit by a neutron bomb. The buildings and public spaces are all there in meticulous detail, but there is no sign of life. The inhabitants and their activities don't show up at all.

This effect results, of course, from a shrewd act of abstraction. Nolli wanted us to see the city as an intricate pattern of architectural masses, public spaces and internally undifferentiated chunks of private space, so he chose a graphic convention that eliminated everything else. He didn't think that what *happened* in the spaces of the city was any of his cartographic business. But it is also true, in any case, that he didn't have the data or the tools to show us more. He had no way to trace the movements of people in detail, to collect and plot statistical data, or to create dynamic displays that revealed flows and transformations.

Mark Imhoff's 1996 satellite-generated city lights maps present similar patterns of black and white, but the abstraction strategy is the inverse of Nolli's.[1] Here, the details of buildings and street patterns disappear completely and you see only the difference between urbanized and non-urbanized areas. Light intensities indicate intensities of human activity. It turns out that these intensities correlate very closely with census data indicating number of

---

[1] earthobservatory.nasa.gov/Study/Lights. Last accessed January 16, 2008.

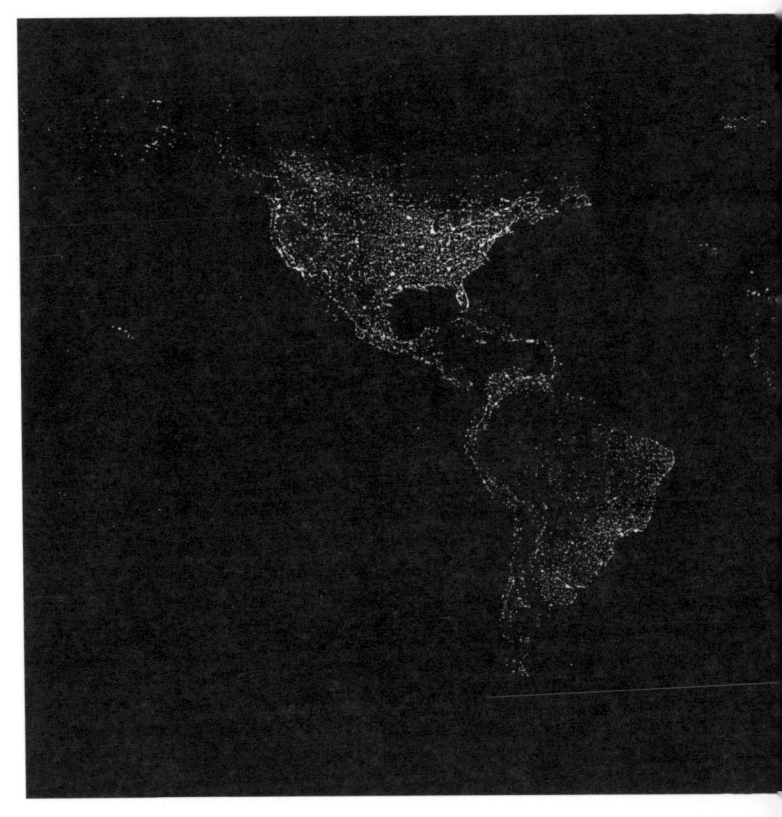

MAPPING URBAN FLUX

Global City Lights, October 23, 2000
credit Data courtesy Marc Imhoff of NASA GSFC and Christopher Elvidge
of NOAA NGDC, image by Craig Mayhew and Robert Simmon, NASA GSFC

people per square mile. So these are maps with no architecture, just traces of the presence of people.

Imhoff's maps are still static; they are snapshots taken at a particular moment and don't show fluctuations of population and activity over time. Furthermore, they don't show people directly, as a close-up aerial snapshot might. To get down to more spatial and temporal detail, it is necessary to tag and track individuals. For good or ill, today's digital electronic technologies now provide the capability to do this with great efficiency.

Automobiles, for example, routinely carry transponders to enable them to pay tolls automatically and enter parking structures. Whenever a transponder is read, the vehicle is identified and its presence at that location is recorded in a log. The data from these logs can be used to track vehicle movements in real time and to create animated displays of dynamically varying traffic flows. Similar RFID (radio frequency identification) technologies can be used to identify, track and map people using cards to enter buildings or public transport systems, pallets and boxes in warehouses, clothing and books in stores, prison inmates, golf balls and just about anything else that can have a tiny, inexpensive electronic tag attached to it.

Increasingly, vehicles also carry GPS (global positioning system) navigation systems. These can be used not only to assist drivers, but also (in combination with wireless reporting capabilities) to compile logs and produce displays of vehicle movements through road networks. The San Francisco Exploratorium's Cabspotting project, for example, produces real time maps showing the movements of Yellow Cabs in the city, together with animations showing changing distributions of cabs over time.[2] These vividly illustrate patterns of daily movement, and when they are overlaid on plots of topographic, demographic and other data, they also begin to suggest explanations for these dynamics.

Until recently, the movements and actions of individual pedestrians in cities could be tracked only in labor-intensive, ex-

---

[2] www.cabspotting.org. Last accessed January 16, 2008.

MAPPING URBAN FLUX

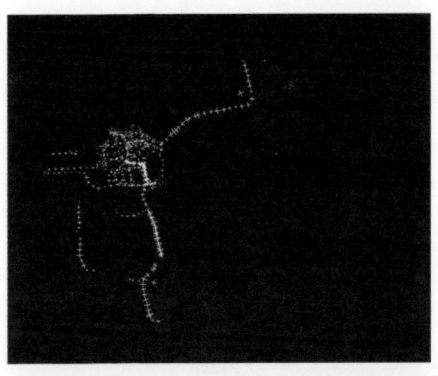

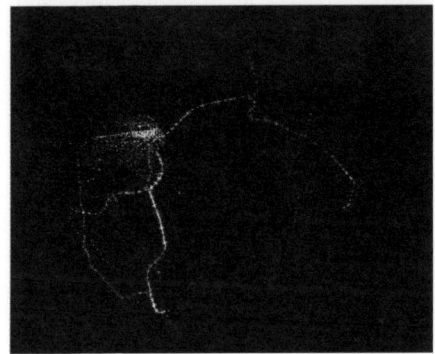

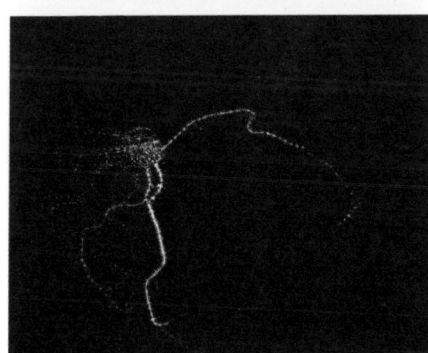

Cabspotting
credit Stamen Design
A project of the San Francisco Exploratorium's Invisible Dynamics initiative. Produced by Peter Richards and Susan Schwartzenberg. Directed by Scott Snibbe with the creative consultation of Amy Balkin. Data generously provided by Yellow Cab of San Francisco. Supported by a grant from the National Foundation for the Arts

pensive ways – for example, by having detectives or paparazzi tail people. Mostly, we could just lose ourselves in the crowd and this was one of the attractions of the big city. Today though, by carrying portable wireless devices – laptop computers and mobile phones, in particular – we have largely given up that anonymity. Whenever we log in at WiFi hotspots, for example, we create entries in system logs. Since hotspots are not very large, these entries precisely locate us in time and space. This provides opportunities for clandestine surveillance of individuals, for optional friend tracking and social networking and for mapping time-varying patterns of space use.[3] And, clearly, it raises social questions that require critical debate.

In addition, whenever we switch on our mobile phones (not just when we make calls) we make our locations electronically discoverable. Our phones can be 'pinged' to determine, at least, the nearest cell tower. Through triangulation of cell tower (or WiFi hotspot) signals, locations can be computed with still greater precision. Advanced phones can also report their GPS coordinates. And when we make calls, we create entries in logs. Since most urban residents carry mobile phones these days, and use them frequently, plotting current wireless traffic intensities on a city map provides a very good picture of where people are in the city.[4] And, through animation of these plots, the rhythms of daily, weekly and seasonal movement can be revealed.

It is now possible to produce even more detailed representations of urban dynamics by considering not just the origins, but also the contents of the signals produced by communications devices. Phone calls, for example, have destinations, and by analyzing destination statistics we can discover the patterns of electronic linkage among locations. Thus, in this exhibit, we can see the linkages of different New York neighborhoods to different parts of the world. The strengths of these linkages fluctuate with time

[3] ispots.mit.edu. Last accessed January 16, 2008.
[4] senseable.mit.edu/grazrealtime. Last accessed January 16, 2008.

of day (largely an effect of time zones), and their patterns reveal a great deal about neighborhood demographics, social behavior, and economic activity.

Technically, it is possible to go much further – for example by mapping Google search statistics or email and telephone content analysis statistics. But the crucial questions that loom here are not ones of technological feasibility. They are questions of how we will want to make some difficult social tradeoffs, how we will debate these tradeoffs, and how much power we still may have to affect them anyway. Who should have the right to track us electronically? When and where? Under what circumstances and controls, and for what purposes? Who should have access to the resulting databases? And in what forms?

In particular, we need to be concerned about levels of data aggregation – a technical-sounding issue, but a socially and politically crucial one. The U.S. Census, for example, collects very detailed and sensitive information about individuals and households, but carefully protects privacy by reporting it only in statistically aggregated ways. Thus we can use publicly available census data to map distributions of household income, for instance – but not down to the level of individual households. Similarly, for billing and other purposes, telecommunications providers collect detailed records of communications traffic, but AT&T supplied only highly aggregated backbone statistics to the New York Talk Exchange project. So the resulting maps and diagrams reveal the patterns of connectivity of neighborhoods and cities, but not of individuals within those geographic units.

There is little doubt that the capacity to track, analyze, and map urban activity in fine-grained, real time detail will open up powerful new ways of providing services to urban inhabitants, of managing urban systems efficiently, of enhancing safety and security, and of making well-informed planning decisions. Consequently, it is clear that government agencies, utilities, and marketers of products and services will continue to press hard to expand the scope and sophistication of their electronic tracking, analysis, and mapping capabilities. But if this expansion does not take place within a framework of vigilant, carefully debated, vigorously formulated and executed public policy – with careful attention to issues of data aggregation and individual privacy – we will soon find that we have unwittingly allowed the stealthy, piecemeal emergence of an irreversible condition of electronic hyper-visibility. We will have allowed individuals, and all that they do, to register on Nolli's map. And we will regret it.

William J. Mitchell is the Alexander Dreyfoos Professor of Architecture and Media Arts and Sciences at MIT and Director of the MIT Design Laboratory. His books include *Imagining MIT*, *Placing Words*, *Me++*, *E-topia*, *City of Bits*, and the forthcoming *World's Greatest Architect*, all from the MIT Press.

ANTHONY M. TOWNSEND AND MITCHELL L. MOSS

NEW YORK: THE CITY OF THE TELEPHONE

Voice communications have been critical to the development of New York City as a center for ideas, information, and culture. Just as the city's ice-free natural harbor led to the rise of trade and commerce, the telephone has shaped New York City's emergence as a global hub for the flow of information in, through and out of the city.

Urban planners often point to Los Angeles as the archetypical automobile city. But New York is the world city most thoroughly organized around the telephone. As economist Robert Lucas has written, "a city, economically, is like the nucleus of an atom."[1] New Yorkers depend on telephones to support the constant flow of data, gossip, and ideas that maintain these bonds.

As long as tools for talking over long distances have existed, New York has been at the forefront of technological adoption. Just as it had played a key role in the early history of the telegraph, what Tom Standage called "the Victorian Internet,"[2] New York was a critical early hub for the emerging telephone infrastructure in the United States in the early

---

[1] Lucas R. E. 2002. *Lectures on Economic Growth* (Harvard University Press: Cambridge, Massachusetts).
[2] Standage T. 1998. *The Victorian Internet: The Remarkable Story of the Telegraph and the Nineteenth Century's Online Pioneers* (Walker & Co.: New York).

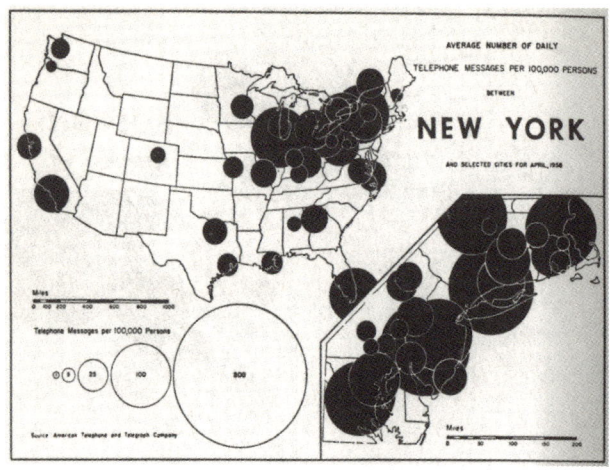

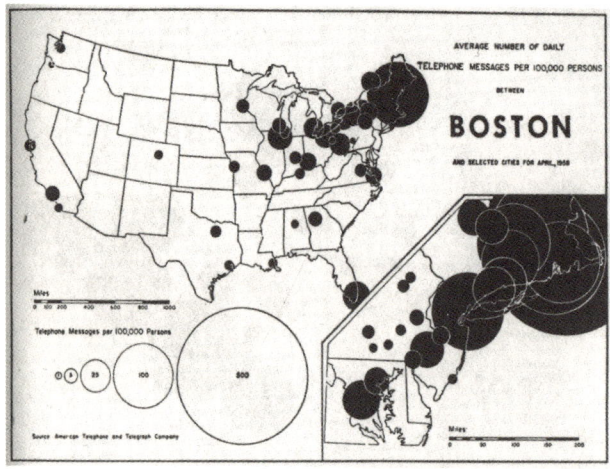

Maps of AT&T phone calls, New York and Boston, April 1958
credit images by Neil C. Gustafson, © 1961 The Twentieth
Century Fund, Inc., reprinted in Gottmann J. 1964. *Megalopolis*
(The MIT Press: Cambridge, Massachusetts)

twentieth-century. As Herbert Casson wrote in an early history of the telephone system:

> Already, by cable, telegraph, and telephone, no two towns in the civilized world are more than one hour apart. We have even girdled the earth with a cablegram in twelve minutes. We have made it possible for any man in New York City to enter into conversation with any other New Yorker in twenty-one seconds.[3]

New York's national and global role would be magnified over the next century as it became connected through the telephone network. The first long-distance telephone link between two cities connected New York and Boston in 1883, and by 1915 transcontinental lines linked the coasts. The telephone became an instrument of cultural and economic domination that continues to this day. As Jean Gottman described in his 1962 studies of "megalopolis" (the urbanized corridor running from Boston to Washington, with New York at its geographic and economic center), New York was a net exporter of information.[4] It is no accident that the original site of Bell Labs, the nation's leading corporate laboratory for telecommunications, was in Manhattan's Greenwich Village.

Today's global telecommunications networks were created in large part because New York's firms were willing to pay for them. The city's knowledge industries – from media firms to investment banks – are powerful drivers of technological innovation. Today, mobile phones are critical to the city's unique role as a vibrant place for face-to-face contact that occurs on the city's sidewalks, in hotel lobbies, office buildings, and restaurants. But it is the complexity of the city's built environment and the density of subscribers that keeps wireless engineers

---

[3] Casson H. 1910. *The History of the Telephone* (Project Gutenberg E-Text: www.gutenberg.org).
[4] Gottman J. 1962. "Megalopolis: The urbanized northeastern seaboard of the United States". *American Political Science Review.*

awake at night, searching for new ways to squeeze more calls into the finite amount of spectrum in midtown Manhattan.

New York was also built on the in-person exchange of information, yet rather than displacing that function, the telephone has enhanced it. The New York Stock Exchange (NYSE) was originally established on a street corner, under a buttonwood tree, where buyers and sellers of stock would meet to conduct business. Today, billions of shares are traded each day on the New York Stock Exchange, through networks that link the NYSE around the world. The telephone has also allowed the rich web of face-to-face transactions that take place in New York to be connected across the globe. Today, the world's major publishing empires are produced by people working in Manhattan but are then distributed electronically to all points of the globe. In a sense, New York imports raw information and exports ideas, decisions and new services.

Surprisingly, despite the importance of the telephone to New York, urban planners know remarkably little about what actually flows across this vital infrastructure. And that's why this exhibition is so timely, because it marks a turning point in our understanding of the city akin to the first aerial photographs.[5] While numerous studies show historical snapshots of telecommunications flows between cities, we are entering an era in which it will become routine to see this data in real time. Measuring information-based activity will be as central to our understanding of cities as the census is today.

We can already see how New Yorkers are using telecommunications to shape globalization. Comparing data provided by AT&T for earlier studies,[6] we see clearly that New York's connections to the world are rapidly diversifying. While the overall

---

[5] For a discussion of the impact of aerial photography on twentieth-century city planning see Campanella T. 2001. *Cities from the Sky* (Princeton Architectural Press: Princeton, New Jersey).

[6] Moss M. L. 1984. "New York Isn't Just New York Anymore". *Intermedia.* 1996 data from unpublished studies of the Taub Urban Research Center, New York University.

Going Global
Percentage of New York City's international telephone traffic (in minutes)

Origin and destination	1982	1996
United Kingdom	23.8	12.1
France	7.5	3.3
Germany	5.9	2.4
Israel	4.2	2.7
Japan	3.3	3.3
Korea	1.4	1.7
China	n/a	1.9
Pakistan	n/a	0.9
Rest of world	53.9	71.7

Source AT&T

volume of telephone traffic expanded several-fold from 1982 to 1996, the share to the city's traditional trading partners in the Western hemisphere declined rapidly. Instead, most of the talk is with new centers of global commerce in Asia and the homelands of the city's exploding immigrant population.

As the twenty-first-century unfolds, New York remains one of the most innovative laboratories for creating new ways of organizing culture and commerce across great distances. The great challenge is to recognize the complex and unpredictable ways in which telecommunications can be used to reinforce the cities of the twenty-first-century. As Casson wrote in 1910:

> No invention has been more timely than the telephone. It arrived at the exact period when it was needed for the organization of great cities and the unification of nations.

Today, the telephone and its many descendants are enabling the organization of new kinds of global structures. The images in this exhibition reveal that our conceptual frameworks have yet to catch up with the reality of information flows. We must be prepared to consider new forms of urbanism, in which communications create communities that transcend spatial boundaries.[7] What we're seeing with these images created by MIT's sense*able* city lab is something much more complex. A city with a center, but few borders. A city with a hinterland inscribed in cyberspace.

Dr. Anthony M. Townsend is a technology forecaster and strategy consultant with the Institute for the Future, based in Palo Alto, California. He consumes massive amounts of bandwidth telecommuting from New York City.

Mitchell L. Moss is the Henry Hart Rice Professor of Urban Policy and Planning at New York University's Wagner Graduate School of Public Service. His research on telecommunications and urban development has been supported by the National Science Foundation and is accessible at www.mitchellmoss.com

---

[7] Sassen S. 1992. *The Global City: New York, London, Tokyo* (Princeton University Press: Princeton, New Jersey).

ALEXANDRE GERBER, MICHAEL MERRITT, CHRIS RATH AND JIM ROWLAND

REPRESENTING
AT&T'S NETWORK DATA IN REAL TIME

The New York Talk Exchange exhibit is a representation of the actual communications traffic flowing to and from New York City via the AT&T network. Data for the project includes the volume of Internet Protocol (IP) traffic flowing through the city from both Internet users and business networks, as well as the number of phone calls occurring in the city at a given time. This data is presented in 10-minute snapshots throughout each day.

The AT&T network today carries 13.4 petabytes of IP traffic on an average business day. That's the equivalent of more than forty-seven megabytes of information for every man, woman and child in the United States. Additionally, its voice network carries 127 billion long-distance voice calls over the course of a year. As the visualizations show, New York City is a center for a large portion of this IP and voice traffic.

For the purposes of this project, both wired and wireless calls are examined. A call is considered to terminate in the city if the telephone switch that serves the dialed number is located within one of the five boroughs of New York. Likewise, a call from a wired phone originates from the city if its phone number is served by one of these switches. Wireless calls are factored in when the cellular tower that serves a phone is located in one of the five boroughs. Thus, calls from New Yorkers who are traveling outside of the city are not counted, because the cellular tower currently serving their phone is outside of the five boroughs. Likewise, wireless calls from visitors to New York City are counted because the cellular tower currently serving them is in one of the five boroughs.

For IP traffic, which can include web browsing, email, business applications, Internet video, gaming or VoIP (voice over IP), volumes were measured as they entered or flowed from the AT&T IP backbone network via one of the network hubs in the tri-state area (New York, New Jersey and Connecticut).

This traffic analysis is performed by researchers at the AT&T Labs Research facilities in Florham Park, New Jersey, using techniques developed to manage traffic flows across one of the world's largest networks to ensure the highest level of service reliability and quality, accurate billing for business and residential customers, and the continued ability to expand and enhance the company's network to meet customers' ever-evolving needs. To realize the New York Talk Exchange project, AT&T Labs Research customized this aggregated traffic data to provide updates in 10-minute intervals to feed the ever-changing display.

AT&T is proud to support New York Talk Exchange, both in collaboration to develop the data that supports the project, as well as sponsorship of the exhibit via the AT&T Foundation.

Alexandre Gerber is a Principal Member of Technical Staff in the Network and Traffic Analysis Research Department at AT&T Labs Research. His research focuses on Internet routing, network management, network data mining, and IP traffic and application performance measurement characterization.

Michael Merritt is Executive Director of the Network and Traffic Analysis Research Department at AT&T Labs Research. He has written more than thirty-five research articles, co-authored a book on database concurrency control, holds four patents and has taught at Columbia University, Georgia Tech, MIT, and Stevens Institute of Technology.

Chris Rath is a computer scientist with AT&T Labs Research. He has worked for AT&T for more than twenty years on a wide variety of projects, including expert systems, text to speech, Internet music downloads, online communities, and cluster computing.

Jim Rowland is currently Director of Data Mining Research at AT&T Labs Research and a former Distinguished Member of Technical Staff at Bell Laboratories. He has focused on incorporating leading-edge research technologies into products and services in areas including artificial intelligence, automatic speech recognition, music compression and data mining.

FRANCESCO CALABRESE

DATA AND SYSTEM ARCHITECTURE

For the New York Talk Exchange project, the MIT sense*able* city lab developed a software platform to collect AT&T's IP and voice call data related to international telecommunications connections between New York City and the rest of the world. The focus of the project was to represent this dynamic data graphically so that visitors to the MoMA exhibition could grasp the richness of this type of global telecommunications activity. In considering which visualizations to compose for the exhibition, we had to think carefully about each data type's characteristics.

Internet Protocol Flows

The data represents the amount of bytes transmitted from and to New York, and is classified either as incoming or outgoing. Incoming data refers to streams of bytes reaching a location in New York from a server located abroad. Outgoing data refers to streams of bytes originating in New York and sent to a server located elsewhere in the world. The IP flows are characterized based on the type of packets exchanged and the type of application involved (such as web browsers, email, peer-to-peer). This allows a differentiation between the kinds of uses of the Internet, such as synchronous communication when there are active participants on all ends and asynchronous communication in the instances when a user connects to a remote server.

The IP flow examined in this project includes all of the Internet traffic going through the AT&T backbone network via one of the network hubs in the tri-state area. The locations at the other end of the stream are derived by geocoding the IP addresses of the packets, which offer a locational accuracy at the city level. The data shows over 35,500 cities from more than 200 countries connecting to New York via AT&T's backbone network.

Voice Call Data

The voice call data represents the amount of long-distance calls received and made from New York City. A call originates and terminates in New York City if a landline or wireless call connects to a switch or base station controller located in one of the five boroughs of the city. The location of these switches is mapped to geographical coordinates with an average accuracy of 500 meters. The location at the other end of the line is calculated by geocoding the country and city code of landline phones and only the country code for wireless phones. The data from AT&T's telephone network reveals that there are more than 27,200 different cities from approximately 250 countries connecting to New York City.

DATA TRANSFER AND VISUALIZATIONS

To feed a secure server at the MIT sense*able* city lab in Cambridge, Massachusetts, AT&T Labs Research customized their aggregated IP and voice traffic data to update every ten minutes. For the MoMA exhibition, the sense*able* city lab server collected the IP and voice call data from AT&T and ran custom-designed software to analyze it and assign the appropriate geographic references. The resulting visualizations were done using Java programming language and Processing software.

The sense*able* city lab servers subsequently sent visualized data to two computers at The Museum of Modern Art in New York, which produced dynamic representations of the telecommunications activity by running different algorithms. Snapshots of the visualizations displayed at the *Design and the Elastic Mind* exhibit are included in the foldout of this book.

Project Credits

The New York Talk Exchange was produced by MIT's sense*able* city lab exclusively for the *Design and the Elastic Mind* exhibition at MoMA, The Museum of Modern Art in New York, New York. The project was shown from February 24 to May 12, 2008 in The International Council of The Museum of Modern Art Gallery, sixth floor. *Design and the Elastic Mind* was organized by Paola Antonelli, Senior Curator, and Patricia Juncosa Vecchierini, Curatorial Assistant, Department of Architecture and Design, The Museum of Modern Art.

The four visualizations were produced using Java programming language and Processing software and were projected onto two separate hanging screens, 84½" by 62¾", at a resolution of 1400 pixels by 1050 pixels.

Project team

Carlo Ratti, sense*able* city lab *director*	Kristian Kloeckl, *project leader*
Assaf Biderman	Francesco Calabrese
Margaret Ellen Haller	Aaron Koblin
Francisca M. Rojas	Andrea Vaccari

Research advisors
William J. Mitchell, Alexander Dreyfoos Professor of Architecture and Media
　Arts and Sciences at MIT and Director of the MIT Design Laboratory
Saskia Sassen, Lynd Professor of Sociology and Member of The Committee
　on Global Thought at Columbia University

Project catalog Clelia Caldesi Valeri

Project sponsor AT&T

Technical partners Yahoo! Design Innovation Team
　　　　　　　　　British Telecom

Acknowledgments

The New York Talk Exchange project would not have been possible without a generous invitation by Paola Antonelli and her team at The Museum of Modern Art to include the sense*able* city lab in their *Design and the Elastic Mind* exhibition. It was also because of Carlo Ratti's audacity and vision to propose an entirely new project for that occasion that the New York Talk Exchange was born.

I want to thank our project sponsor AT&T, in particular Alexandre Gerber, Michael Merritt, Chris Rath and Jim Rowland of AT&T Labs who have been partners in this project in the true sense of the word. Without their dedication, patience and effort in elaborating and structuring the data, this ambitious project of data visualization would have been impossible to accomplish. Marilyn Reznick of the AT&T Foundation also played a fundamental role in facilitating the necessary contacts and circumstances needed for such a fertile collaboration between AT&T and the sense*able* city lab to flourish.

Throughout this project we have had the pleasure of engaging the notable academics Saskia Sassen and William J. Mitchell as research advisors. They have been invaluable in bolstering the project's strengths and tackling its weaknesses in order to put this complex project on solid ground. The richness and depth of their contribution is evident in the diversity of issues addressed in their essays. We are also thankful that the urban telecommunications scholars (and New Yorkers) Anthony Townsend and Mitchell L. Moss contributed their illuminating historical perspective to this project.

Remarkably, the New York Talk Exchange was organized and assembled in three short months prior to the opening of the MoMA exhibition. Development steps which are usually taken sequentially were turned into parallel processes that went hand in hand: while the AT&T data streamed into our server and was transformed into visual expression, the patterns that emerged

Printed by
Aktiva - corso Moncalieri 448 - 10133, Turin, Italy

Photolitho by
Fotomec in Turin, Italy

were fed back into the structuring and elaboration of the data itself, and so on, in numerous rounds of refinement.

Realizing this project has been possible entirely because of the project team's wonderful dedication and determination at the sense*able* city lab. It is a true interdisciplinary group in which computer scientists contribute also to the process of visual structuring and content while designers, urbanists and theorists tackle the raw data to figure out how to make it speak.

I extend my gratitude to the entire team:

Francisca Rojas for considering how urban theory and the AT&T data together could open up new perspectives on the city and how this could be best communicated to an audience. Francesco Calabrese who masterminded the system architecture and whose overall commitment and support to every single aspect of the project has been felt throughout. Aaron Koblin, whose imaginative contributions have been determinant in this data-driven project of painting with pixels. Andrea Vaccari, for insisting on making the world-map twist and the data flow in real time. Margaret Ellen Haller, who made sure that MoMA and our team were happily on the same page. And Assaf Biderman, for his welcomed insights into directions we were taking and others that could be explored.

Last but not least, I am grateful to Clelia Caldesi Valeri for realizing this beautiful book after having developed a concept that allowed the content to be elaborated concurrently with the final stages of the project's completion.

I am certain that this exhibition project will be the beginning of a promising and continuing research agenda.

Kristian Kloeckl
Cambridge, Massachusetts
January 28, 2008